New Zealand Society of Animal Production Conference (NZSAP 2023)

The land, the water, the people.

Proceedings of the New Zealand Society of Animal Production
Volume 83

Rotorua, New Zealand
14 - 16 November 2023

ISBN: 978-1-7138-9812-2

Printed from e-media with permission by:

Curran Associates, Inc.
57 Morehouse Lane
Red Hook, NY 12571

proceedings.com

Some format issues inherent in the e-media version may also appear in this print version.

New Zealand Society of Animal Production (Inc.) (NZSAP)
c/o Kate Crookston
PO Box 955
Cambridge, New Zealand

www.nzsap.org.nz

Additional copies of this publication are available from:

Curran Associates, Inc.
57 Morehouse Lane
Red Hook, NY 12571 USA
Phone: 845-758-0400
Fax: 845-758-2633
Email: curran@proceedings.com
Web: www.proceedings.com

TABLE OF CONTENTS

BRIEF COMMUNICATION: An evaluation of the AGTECH360 Livestock Tracker for the automated recording of on-farm grazing events.

WA Hofmann[1*], JP Edwards[2] & MB Neal[1]

[1]*DairyNZ Ltd, Private Bag 3221, Hamilton 3240, New Zealand,* [2]*DairyNZ Ltd, Millpond Lane, Lincoln, 7608, New Zealand*

*Corresponding author. Email: Wayne.Hofmann@dairynz.co.nz

Abstract

Many New Zealand farmers either do not record paddock grazing events or do not retain grazing data after making decisions about cow feeding and pasture management. This study, conducted at the Northland Agriculture Research Farm (NARF), aimed to determine whether GPS-enabled devices suitable for use on cattle could be used to automate the recording of grazing events on-farm. Accordingly, 12-13 cows within each of the three herds (herd size range 59-85 cows) were fitted with an AGTECH360 Livestock Tracker from 1 December 2021 to 31 March 2022. The GPS location data were then compared with manual grazing records. The grazed paddock was correctly identified on 80-92% of occasions, depending on the herd. These results indicate that it is feasible to automate the recording of on-farm grazing events by fitting a proportion of the herd with GPS-enabled devices.

Keywords: dairy farm systems; grazing records; global positioning systems (GPS); Internet of Things (IoT); low-power wide-area network (LPWAN); LoRa

Introduction

Grazed pasture is the primary resource that underpins dairy production and farm profitability in countries with temperate climates, such as New Zealand (Hanrahan et al. 2018; Neal & Roche 2019). In these climates, year-round pasture growth is achievable, albeit in a seasonal pattern, making pasture a low-cost, high-quality feed source if well managed. This can reduce reliance on purchased feeds with a higher cost and environmental footprint (Belflower 2010; Rotz et al. 2020). High on-farm pasture utilisation requires information on individual paddock performance and grazing records. However, while the recording of grazing events and pasture measurement are essential parts of pasture management, they are often neglected due to their time-consuming nature (Anderson & McNaughton 2018). While some farmers conduct regular farm walks and record pasture covers, this information is typically only used for short-term feeding and farm management decisions (i.e., where the cows will graze today) and not retained once these decisions have been completed (Stevens & Knowles 2011).

The recent expansion of the Internet of Things (IoT) has led to the development of new low-power wide-area networks (LPWAN), such as Long Range (LoRa) radio communication techniques (Augustin et al. 2016), which has led to an increase in the availability of Global Positioning System (GPS) devices suitable for use on cattle. Subsequently, there has been renewed interest in using GPS devices in the dairy sector. For example, they could be used to automate the recording of on-farm grazing events (Haultain 2014; Hofmann 2022; Hofmann et al. 2022). Alternatively, Woodward et al. (2019) incorporated GPS cow data and spectral data to estimate pasture mass. Our research evaluated the feasibility of automating the recording of paddocks grazed on-farm by fitting a proportion of the dairy herd with the AGTECH360 Livestock Tracker (AGTECH360 2020, Queensland, Australia).

Materials and methods

Device

The AGTECH360 Livestock Tracker is a reusable, lightweight (approximately 32 g), solar-powered ear tag measuring approximately 5.5 cm (wide) × 7 cm (high) with an estimated five-year lifespan (AGTECH360 2020; Meat and Livestock Australia 2019). The devices use a LoRa network to transfer data from the device to a base station and a cellular network to transfer data from the base station to the cloud and were configured to obtain a GPS position approximately hourly (24 daily positions) per the manufacturer's default setting. Previously, Hofmann (2022) conducted static testing of these devices at Scott Farm, Newstead, New Zealand (-37.769° S, 175.366° E) and reported a mean location error, which is the difference between an object's actual position and that estimated by a GPS position fix, of 5.4 m and a 95% Circular Error Probability (CEP) of 13.9 m. The CEP is a circular radius containing a stated percentile of points around an actual location (GPS device's proper location) (Morris & Conner 2017; Turner et al. 2000)

Research site

On-farm testing occurred at the Northland Agricultural Research Farm (NARF), Dargaville, New Zealand (-35.943° S, 173.841° E) between 1 December 2021 and 31 March 2022. This is an 82-hectare (effective) dairy farm. As part of the four-year Future Farm Systems Trial established on 1 June 2021 (Northland Dairy

Development Trust 2023), the farm is divided into three farmlets of approximately 27 hectares each, with paddock sizes ranging from 0.80 to 1.75 ha. Over the study period, the cows were milked twice daily. Herd sizes were 59 cows for the green farmlet, a low-stocked farmlet aimed at reducing greenhouse gas emissions, 83 for the blue farmlet, a farmlet testing alternative pasture species and 85 for the red farmlet, operating under a typical Northland farming system.

Data collection

Twelve cows from the green farmlet and 13 cows each from the red and blue farmlets (~14-22% of the milking herd) were randomly selected and fitted with an AGTECH360 Livestock Tracker. The Lincoln University Ethics Committee (AEC 2021-12) approved the application of devices. Throughout the trial period, the devices remained fitted to the same cows irrespective of whether they remained in the milking herd or were drafted out for treatment or other purposes. The LoRa base station was installed at the dairy for easy access. Essential information recorded by the devices included location (latitude and longitude), date and time of observation, and ambient air temperature. On most occasions, the milking herds were offered a proportion of a paddock with separate night and day paddocks. Daily grazing records (i.e., date, paddock number, percentage paddock allocated, and cow numbers) for each farmlet were recorded manually by NARF staff using a Microsoft Excel spreadsheet.

A digital map of the property (Figure 1) was produced in RStudio (version 4.2.1; R Core Team 2022) using the dplyr and sf packages in R (version 1.0.12; Pebesma 2018; Pebesma & Bivand 2023; R Core Team 2022). This made it possible to interpret the GPS data collected by the devices in terms of paddocks grazed over the trial period. The spatial join function st_join was used to identify which GPS points were recorded in which paddocks or in other farm areas, e.g., at the cowshed or on farm laneways. The GPS points were then grouped into periods corresponding to the grazing area for each farmlet, based on the farm milking times of 6 am and 3 pm (e.g., between 6 am and 3 pm, two grazing periods were applied: 6:00 to 10:30 am and 10:30 am to 3:00 pm). Multiple grazing periods were applied between milkings as, on occasions, several paddocks were grazed between milkings depending on the farmlet's needs, e.g., using a sacrifice paddock for the blue farmlet to protect the pastures on that farmlet from overgrazing. Additionally, the pm grazing was divided into two periods (3:00 pm to 00:00 am and 00:00 to 6:00 am) to ensure the GPS record date was correctly aligned with the manually recorded grazing dates. The paddock with the highest number of GPS observations in each grazing period was then selected as the most likely paddock grazed by the herd during each grazing period. Subsequently, the GPS grazing records were compared with the manual on-farm grazing records to identify which grazing areas the GPS devices correctly identified.

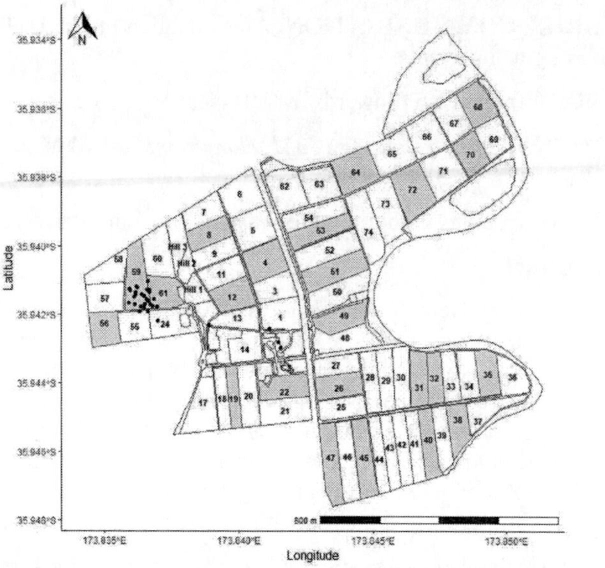

Figure 1 An Outline of the NARF and the GPS fixes recorded by 13 AGTECH360 Livestock Trackers on the green farmlet (shaded paddocks) for one grazing period (01 January 2022) between 6:00 and 10:30 am. Cows were allocated the first one-third of paddock 59 for this grazing period.

Results and discussion

The paddock grazed by the dairy herd was correctly identified on 79.6% of occasions for the blue farmlet, 91.3% for the red farmlet and 91.9% for the green farmlet over the four-month study period (Table 1).

There are several reasons why a 100% match rate between the GPS observations and the on-farm grazing records was not achieved. Firstly, GPS positions were sometimes either not reported, or the ear tags reported intermittently to the on-farm LoRa base station when grazing a particular farm area, possibly due to the positioning of the base station relative to these paddocks on-farm. Secondly, the blue farmlet cows spent time in sacrifice paddocks in February and March. However, these grazing events, which comprised 16.6% of the total grazings for this farmlet, were not included in the manual record; hence, the lower percentage of grazings correctly matched for this farmlet. If these grazing events are excluded from the GPS records, the result aligns with the remaining two farmlets. Thirdly, when solely selecting the paddock with the highest number of GPS observations during each grazing period, the incorrect paddock was identified in the GPS records for 3.8% of grazing events for the blue farmlet, 8.1% for the green farmlet, and 8.7% for the red farmlet which may be due to paddock shape or size or device accuracy. Finally, on occasions, there were equal numbers of GPS points in two or more paddocks within the same grazing period, occurring in < 1% of grazing periods for the blue farmlet and in 1.5% of grazing periods for the red and green farmlets.

Table 1 Key summary statistics from 1 December 2021 to 31 March 2022.

Farmlet	Herd Size	Number of Tags	Percentage of Herd Tagged	Number of Paddocks Grazed – NARF Records	Number of Paddocks Grazed – GPS Estimate	Percentage of Paddocks Grazed Correctly Assigned
Red	85	12	14.1	220	231	91.3
Green	59	13	22.0	221	234	91.9
Blue	83	13	15.7	236	289	79.6

In the four-month testing period (1 December 2021 to 31 March 2022), two tags or 5.3% of tags applied (one each from the green and blue farmlets) stopped reporting daily GPS positions, possibly due to tag failure or loss. Tag losses were not recorded. Across all farmlets, the devices reported an average of 19.6 positions per day throughout the trial period, close to their programmed rate of 24 daily positions.

One of the main factors that is likely to influence farmers' adoption of this technology is cost, with each AGTECH360 device costing around NZ$100 per tag in addition to the base-station cost of NZ$2000. Previous studies (Haultain 2014; Hofmann 2022; McGranahan et al. 2018) have suggested that a minimum of 1% of cows, or at least three cows per herd, should be fitted with GPS devices to record on-farm grazing events automatically. This number is lower than the current study, which used 12-13 devices per herd. The higher number of devices likely contributed to the high proportion of correctly identified paddocks in this study (79.6% or more). An earlier study by Hofmann et al. (2022) reported that in an on-farm study with 2.75% of the herd tagged with AGTECH360 devices, 52.5% of all recorded observations were within the correctly allocated paddock (i.e., the paddock the cows were grazing). To achieve high accuracy of grazing events, more than 1% of animals must be tagged based on the results of this study and previous studies (Haultain 2014; Hofmann 2022; Hofmann et al. 2022).

An important consideration when selecting GPS devices to identify the grazing area is the device's fix rate, which is how often a GPS position is recorded. A more frequent fix rate (i.e., every 30 minutes vs. once every hour) will improve the accuracy of identifying the grazing area (Swain et al. 2008; Turner et al. 2000). However, the fix rate cannot be selected in isolation. For example, a higher fix rate will increase the battery demand of the devices. Therefore, the end user will likely need to reach a compromise between several factors, including the fix rate, the number of devices required, and the error rate they are willing to accept.

Accurately recording paddock grazings provides valuable insights for management, including identifying which paddocks are growing best (e.g., grazed more often), and allows farmers to track rotation length or prepare a paddock shortlist to assess visually. Grazing data is also invaluable in feed planning and farm management decisions, such as pasture renewal and fertiliser usage. Automated grazing records could also be inputs to farm management software such as Pasture Coach or Minda Land and Feed (Hammond 2017).

These study results indicate that using GPS-enabled devices to automate the recording of on-farm grazing events is feasible, potentially leading to improvements in on-farm pasture management. However, further work is required before wide-scale use becomes possible. Research is currently underway investigating alternative GPS devices, including direct-to-satellite solar-powered ear tag options such as the Ceres Trace (Ceres Tag 2023) and GSatRancher (Global Satellite Engineering 2023), that may be able to overcome the coverage and set up constraints of a LoRa system.

Acknowledgements

Data for this study were collected as part of the New Zealand Bioeconomy in the Digital Age (NZBIDA) programme funded by AgResearch and New Zealand dairy farmers through DairyNZ Inc. and in-kind support from Fonterra Co-operative Group Limited. The authors wish to acknowledge the staff and management of the NARF for allowing this work to take place and for assisting with recording grazing events and tag application.

References

Agtech360. 2020. Farm asset tracker range [Online]. Available: https://agtech360.net/asset-trackers/ [Accessed 1 May 2023].

Anderson G, McNaughton L 2018. Validation of a satellite pasture measurement system. Proceedings of the 8th Australasian Dairy Science Symposium, 2018 Palmerston North, New Zealand. 191-195.

Augustin A, Yi J, Clausen T, Townsley WM 2016. A study of LoRa: Long range and low power networks for the internet of things. Sensors, 16 (9): 1466. https://doi.org/10.3390/s16091466

Belflower J. 2010. Environmental assessment of pasture-based and confined dairy farms in Georgia. Masters thesis, University of Georgia, Athens, Georgia. https://getd.libs.uga.edu/pdfs/belflower_jefferson_b_201012_ms.pdf

Ceres Tag. 2023. About us [Online]. Available: https://cerestag.com/pages/about-us [Accessed 25 August 2023].

Global Satellite Engineering. 2023. The GSatSolar Series [Online]. Available: https://www.gsatsolar.com/ [Accessed 25 August 2023].

Hammond H 2017. Agricultural software: A case study of feed and animal information systems in the New Zealand dairy industry. Masters thesis, Massey University, Palmerston North, New Zealand. https://mro.massey.ac.nz/handle/10179/12501

Hanrahan L, McHugh N, Hennessy T, Moran B, Kearney R, Wallace M, Shalloo L 2018. Factors associated with profitability in pasture-based systems of milk production. Journal of Dairy Science, 101 (6): 5474-5485. https://doi.org/10.3168/jds.2017-13223

Haultain J 2014. Ranking paddock performance using data automatically collected in a New Zealand dairy farm milking system. Masters thesis, Massey University, Palmerston North, New Zealand, https://mro.massey.ac.nz/handle/10179/5423

Hofmann WA 2022. An evaluation of GPS technology as a tool to aid pasture management. Masters thesis, University of Waikato, Hamilton, New Zealand. https://researchcommons.waikato.ac.nz/handle/10289/14864

Hofmann WA, Neal MB, Woodward SJR, O'Neill TA 2022. GPS technology as a tool to aid pasture management on dairy farms. Journal of New Zealand Grasslands, 84: 189-196. https://doi.org/10.33584/jnzg.2022.84.3561

McGranahan DA, Geaumont B, Spiess JW 2018. Assessment of a livestock GPS collar based on an open-source datalogger informs best practices for logging intensity. Ecology and Evolution, 8 (11): 5649-5660. https://doi.org/10.1002/ece3.4094

Meat and Livestock Australia. 2019. Smart tracker on the move [Online]. Available: https://www.mla.com.au/news-and-events/industry-news/smart-tracker-on-the-move/ [Accessed 28 April 2023].

Morris G, Conner LM 2017. Assessment of accuracy, fix success rate, and use of estimated horizontal position error (EHPE) to filter inaccurate data collected by a common commercially available GPS logger. PLOS One, 12 (11). https://doi.org/10.1371/journal.pone.0189020

Neal M, Roche JR 2019. Profitable and resilient pasture-based dairy farm businesses in New Zealand. Animal Production Science, 60 (1): 169-174. https://doi.org/10.1071/AN18572

Northland Dairy Development Trust. 2023. Current trials [Online]. Available: https://nddt.nz/current-trials/ [Accessed 10 May 2023].

Pebesma E 2018. Simple Features for R: Standardised support for spatial vector data. The R Journal, 10 (1): 439-446. https://doi.org/10.32614/RJ-2018-009

Pebesma, E. Bivand R 2023. Spatial data science: With applications in R (1st ed.), New York, Chapman and Hall/CRC. https://doi.org/10.1201/9780429459016

R Core Team. 2022. R: A language and environment for statistical computing. R Foundation for Statistical Computing, Vienna, Austria. [Online]. Available: https://www.R-project.org/ [Accessed 1 May 2023].

Rotz CA, Holly M, De Long A, Egan F, Kleinman PJ 2020. An environmental assessment of grass-based dairy production in the northeastern United States. Agricultural Systems, 184. https://doi.org/10.1016/j.agsy.2020.102887

Stevens D, Knowles I 2011. Identifying the need for pasture renewal and valuing the contribution of renewal on a dairy farm-Telford Dairy, a case study. NZGA: Research and Practice Series, 15: 211-216.

Swain D, Wark T, Bishop-Hurley G 2008. Using high fix rate GPS data to determine the relationships between fix rate, prediction errors and patch selection. Ecological Modelling, 212 (3-4): 273-279.

Turner L, Udal M, Larson B, Shearer S 2000. Monitoring cattle behaviour and pasture use with GPS and GIS. Canadian Journal of Animal Science, 80 (3): 405-413.

Woodward SJR, Neal MB, Cross PS 2019. Preliminary investigation into the feasibility of combining satellite and GPS data to identify pasture growth and grazing. Journal of New Zealand Grasslands, 81: 47-54. https://doi.org/10.33584/jnzg.2019.81.404

BRIEF COMMUNICATION: Effects of the SLICK gene on heat tolerance in grazing dairy cattle

GM Worth, EG Donkersloot, RJ Spelman & SR Davis*

Livestock Improvement Corporation Ltd., Newstead, Hamilton 3240, New Zealand

**Corresponding Author:* Email steve.davis@lic.co.nz

Abstract

The heat tolerance of Senepol beef cattle (Bos taurus) has been causally linked to a mutation in the prolactin receptor (PRLR) on BTA20. This variant of PRLR has become known as the SLICK gene because it is also associated with a short-hair coat. We have investigated the heat tolerance of New Zealand dairy cattle bred to carry the SLICK gene, using measurement of vaginal temperature. Hair length differed (P<0.001) between SLICKs and controls at the neck (4.3 ± 0.5 mm vs 13.8 ± 1.2 mm, respectively), shoulder (6.5 ± 0.7 mm vs 14.9 ± 1.5 mm, respectively) and pin bone (6.9 ± 0.6 mm vs 15,2 ±1.4 mm, respectively). Vaginal temperatures of SLICK animals were 0.5 - 1°C lower than controls at peak ambient temperature (28.4°C; 16.00h). SLICK dairy cattle show a markedly improved ability for temperature homeostasis under mild heat load, relative to control cattle.

Keywords: heat tolerance; dairy cattle; SLICK; genetics

Introduction

Heat stress has significant welfare implications for dairy cattle, impacting dry matter intake, daily milk production, calf birthweight, cow fertility and other important traits (Garner et al. 2017). Climate change is driving a need to improve heat tolerance of the New Zealand (NZ) dairy population to protect animal welfare and maintain milk production. The NZ pastoral system does not allow, in practical terms, for use of many of the cooling systems available within housed systems, which puts increased emphasis on finding alternative, including genetic, solutions.

The Senepol breed of cattle, a beef breed that originates from the Caribbean, is unusual among Bos taurus breeds in that it is heat tolerant. The heat tolerance has been causally linked to a mutation in the prolactin receptor (PRLR) on BTA 20 (Littlejohn et al. 2014). This variant of PRLR has become known as the "SLICK gene" because it is also associated with a short-hair coat. Both the SLICK coat and heat tolerance are dominant traits.

Since 2014, a breeding programme has been underway at LIC to introgress the SLICK variant into elite NZ dairy cattle. In addition to enhancing the genetic merit of SLICK offspring, studies are underway evaluating the heat tolerance of NZ dairy cattle carrying the SLICK gene with control cattle.

Materials and methods

The trial was conducted January/February 2021 at the LIC Innovation farm, Hamilton. All procedures involving animals were approved by the Ruakura Animal Ethics Committee (Approval Number 15178) under the New Zealand Animal Welfare Act 1999.

Nine, 2018 born, heterozygous SLICK heifers (12.5% Senepol 87.5% NZ Holstein Friesian/ Jersey crossbred) calved in August 2020 and were studied as part of LIC's commercial milking herd during the second lactation. The SLICK heifers were matched (age, size and coat colour) with a control group of nine Friesian-Jersey crossbred heifers. Animals remained within the herd for the duration of the trial and were milked twice a day. Vaginal temperature was recorded every 10 min by Star Oddi DST Centi-T temperature loggers (Star-Oddi, Gardabaer, Iceland). Loggers were attached to a shortened cow CIDR (hormone-free; Zoetis, New Zealand) with 2 layers of heat-shrink and inserted into the vagina for 7 days (Schutz et al. 2023). Hair samples were collected in summer (January) by taking a finger-pinch of hairs from 3 regions (pins, shoulder and neck) for length measurement using ImageJ software. A total of 15 hairs were measured from each region sampled, for all animals. Air temperature (°C), and relative humidity (%),were recorded at 10-min intervals (Wireless Vantage Pro2™ Integrated Sensor Suite, model 6322, Davis Instruments Hayward, CA, USA) and the temperature-humidity index (**THI**; Igono et al., 1992) calculated. THI is an estimate of the environmental impact on the cow's ability to lose heat by convection and evaporation. However, THI does not include solar radiation load and windspeed which are important parameters in the environment of grazing cattle.

Results

The mean vaginal temperature was similar in both groups in the morning on both days of observation (~38.3°C; Fig 1) with a small increase seen around morning milking (~5.00h). Vaginal temperature was similar in both SLICK and control groups on the hot day (THI 76 at 15.00h) until around 12.00h. Between 14.00 and 20.00h, vaginal temperature in the SLICK cows was relatively constant at ~38.5°C while the mean vaginal temperature in the control group rose to a peak of 39.5°C, declining back to 38.5°C by 20.00h (Fig 1). The temperature increase can be associated, in part, with the heat generated by walking to the shed for afternoon milking (~14.30h). Vaginal temperature among the SLICK animals was less variable than the controls, particularly during the afternoon when temperatures were at their highest. Vaginal temperatures observed on the cooler day (Fig. 1, 1b, THI 67 at 15.00h)) remained relatively constant for both groups, with just a small increase in vaginal temperature observed in both groups in late afternoon.

Figure 1 Mean vaginal temperature for SLICK (black) and control (grey) cows on two days, where the THI and ambient temperature (dotted line) at 15.00h was: 1a, 76 and 28.4°C, 1b, 67 and 21.7°C. Standard error bars are shown.

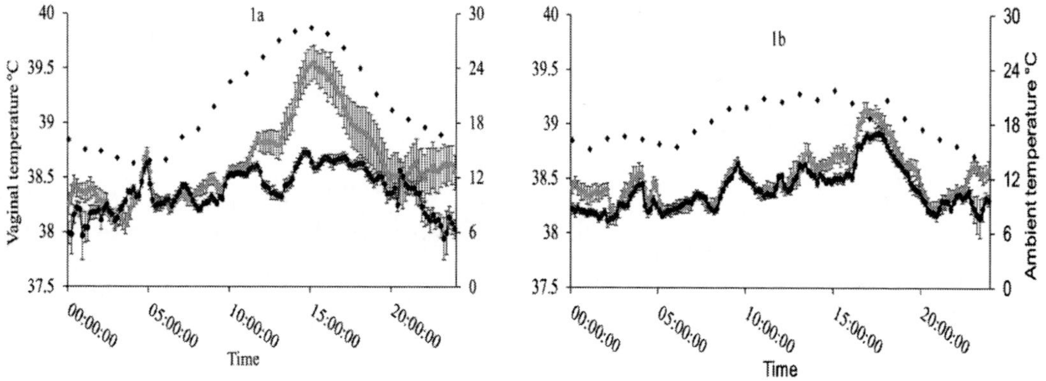

Hair samples collected and measured from all three locations were significantly shorter ($P<0.001$) in the SLICK cattle than controls. For both groups, hair length was shortest on the neck and longest on the pin (Table 1).

Table 1 Mean hair length measurements (± SEM) collected from SLICK and control cattle in mid-summer.

	Hair length (mm)		
Location	Control	SLICK	P value*
Neck	13.8 ± 1.16	4.3 ± 0.46	<0.001
Shoulder	14.9 ± 1.50	6.5 ± 0.70	<0.001
Pin	15.2 ± 1.42	6.9 ± 0.61	<0.001

***By Student's 't' test.** JMP (Version 8, SAS Institute Inc., Cary, NC)

Discussion

The results from this study are consistent with previous findings that under a relatively mild, summer heat load, SLICK cattle have a markedly improved ability for temperature homeostasis (Dikmen et al. 2014), associated with substantially shorter hair length. Several physiological factors may be involved with SLICK cattle's improved ability to regulate body temperature. The short summer coat observed in SLICK cattle should aid in increasing the rate of heat loss via convection and conduction (Berman, 2004). Further, Dikmen et al. (2014) observed that the SLICK trait was also associated with an increased sweating rate. Respiration rate has also been reported to be lower in SLICK relative to control dairy cows when under heat load (Dikmen et al. 2014), a response we have also observed (data not shown).

Further investigation of the mechanism of action of the SLICK variant/prolactin pathway is needed. Improved heat dissipation is likely the primary mechanism of SLICK as limited data shows that the SLICK variant allows dairy cattle to maintain a higher level of production when under heat stress (Dikmen et al. 2014). Hair length in heat tolerant, Bos indicus breeds is even shorter than that seen in SLICK cattle (Davis et al. 2017) contributing to the well-known, heat tolerance of zebu breeds. Definition of the role of the SLICK coat in enabling heat dissipation is needed, including the impact of the SLICK variant on the characteristics of the winter coat.

However, also of interest is the greater variability in heat tolerance that is seen among the control cows (Fig. 1a). Some of these animals had similar heat tolerance ability to the SLICK group indicating that, as might be expected, there are other genetic influences on heat tolerance in NZ cattle that have yet to be identified. Nevertheless, introgression of the SLICK gene into the dairy cattle population in New Zealand has the potential to substantially improve heat tolerance and enhance productivity during heatwaves, which will assist in mitigation of the predicted effects of climate change.

Acknowledgements

The authors gratefully acknowledge the assistance of Joel Riwhi and Rowan Parkin at the LIC Innovation farm, Rukuhia, Hamilton.

References

Berman A 2004. Tissue and external insulation estimates and their effects on prediction of energy requirements and of heat stress. Journal of Dairy Science 87: 1400-1412.

Davis SR, Spelman RJ, Littlejohn MD 2017. Breeding and Genetics Symposium: Breeding heat tolerant dairy cattle: the case for introgression of the "slick" prolactin receptor variant into Bos taurus dairy breeds. Journal of Animal Science 95: 1788-800.

Dikmen SE, Khan FA, Huson HJ, Sonstegard TS, Moss JI, Dahl GE, Hansen PJ 2014. The SLICK hair locus derived from Senepol cattle confers thermotolerance to intensively managed lactating Holstein cows. Journal of Dairy Science 97: 5508-20.

Garner JB, Douglas M, Williams SR, Wales WJ, Marett LC, DiGiacomo K, Leury BJ, Hayes BJ 2017. Responses of dairy cows to short-term heat stress in controlled-climate chambers. Animal Production Science 57: 1233-41.

Igono MO, Bjotvedt G, Sanford-Crane HT 1992. Environmental profile and critical temperature effects on milk production of Holstein cows in desert climate. International Journal of Biometeorology 6: 77-87.

Littlejohn MD, Henty KM, Tiplady K, Johnson T, Harland C, Lopdell T, Sherlock RG, Li W, Lukefahr SD, Shanks BC, Garrick DJ et al. 2014. Functionally reciprocal mutations of the prolactin signalling pathway define hairy and SLICK cattle. Nature Communications 5(1): 5861.

Schütz KE, Cox NR, Cave VM, Huddart FJ, Tucker CB 2023. Effects of changing milking and feeding times on the behaviour, body temperature, respiration rate and milk production of dairy cows on pasture. Applied Animal Behaviour Science 261: 105895.

New Zealand Journal of Animal Science and Production 2023 Vol 83.

BRIEF COMMUNICATION: Approaches to breeding heat tolerant dairy cattle

EG Donkersloot, GM Worth, LR McNaughton, A Yeates, SR Davis* & RJ Spelman
Livestock Improvement Corporation, Private Bag 3016, Hamilton 3240, New Zealand
Corresponding author:* Email **steve.davis@lic.co.nz

Abstract

Climate change will become increasingly problematic for grazing cattle, with heat stress impacting both animal welfare and productivity. Genetic strategies are available to mitigate heat stress, some of which such as crossbreeding, have already been tested in tropical climates. Heat tolerance is a heritable trait in dairy cattle and genetic progress can be achieved using genomic selection, an approach adopted in Australia. Another strategy is to introgress the SLICK genetic variant from Senepol beef cattle into dairy cattle. This enables a 0.5 to 1.0 °C advantage in rectal and ruminal temperature, relative to control cows, when under a mild heat load. This variant has already been introduced into breeds of beef cattle by gene editing, an approach which promises to greatly accelerate gene introgression for single gene traits and lessen the potential impact of introgression on inbreeding in a population.

Keywords: heat tolerance, dairy cattle, SLICK, breeding

Introduction

Heat stress, and the accompanying loss of milk production in dairy cattle, begins when environmental temperatures start to exceed 20 °C (Bryant et al., 2007). In New Zealand, the number of days in the year when cows are subject to heat stress is expected to increase with climate change, thus accelerating the need to produce management and genetic tools to mitigate the effects of this more challenging environment (Harburg et al., 2019).

The accumulation of heat in the dairy cow leads to heat stress when the animal is unable to dissipate heat at a faster rate than heat loading. Heat stress will manifest itself as increased body temperature and increased respiration rate as well as multiple physiological sequalae, including reduced feed intake, fertility and milk production (Garner et al. 2017).

Heat tolerance is a heritable trait in dairy cattle (eg Nguyen et al. 2016). The least heat tolerant animals are large, high-producing, Holstein (Bos taurus) cattle and the most heat tolerant are the less productive and smaller zebu-type (Bos indicus) cattle. This paper discusses some of the historical and future breeding options to improve the heat tolerance trait in cattle with some focus on the potential of the "SLICK" genetic variant, found in Criollo cattle breeds from the Caribbean, that is associated with production of a short hair coat and improved heat tolerance.

Cross breeding

In tropical climates, one breeding strategy to combat heat stress has been to introduce improved temperate genetics (often Holstein) into local Zebu (Bos indicus) breeds to improve lactation performance while retaining necessary traits such as heat tolerance. This strategy has never been totally satisfactory, in part because of the need to provide improved, and sometimes uneconomic, feeds to support milking performance (McDowell, 1985). However, composite breeds eg Girolando, have been developed which have successfully combined the desired attributes of Holstein and Zebu (Gir) and are in commercial use in South America (Madalena, 1998).

Selection for heat tolerance within a dairy population

Heat tolerance is a polygenic trait and is therefore amenable to a selection approach provided a suitable phenotype(s) can be measured. Nguyen et al. (2016) used the decline in milk, fat and protein as proxies for heat tolerance and developed a genomic prediction which had sufficient accuracy to support the development of a heat tolerance breeding value for dairy cattle in Australia (Pryce et al. 2023). Similar initiatives are not being undertaken, currently, in NZ.

The breeding values produced were negatively correlated with production but positively correlated with fertility (Nguyen et al. 2016). In New Zealand, Bryant et al. (2007) employed a similar phenotyping approach which indicated that production loss in Friesians began at a temperature-humidity index (THI) of 68 whereas in Jerseys, production loss began at a THI of 75. However, in this dataset it was impossible to separate the direct production loss effects from the potential impacts of environmental temperature on pasture quality.

Introgression of the "SLICK" variant into dairy breeds

Olson et al. (2003) demonstrated that a major gene for coat length and heat tolerance was segregating in Senepol beef cattle (Bos taurus) ultimately leading to the discovery of the "SLICK" genetic variant (a dominant, major gene) located in the prolactin receptor (PRLR; Littlejohn et al. 2014). Since the initial discovery, a family of variants in the PRLR gene have been identified in Criollo breeds in the Caribbean that elicit the same changes in coat type and heat tolerance (Porto-Neto et al. 2018). All of these variants truncate the cytoplasmic domain of PRLR.

A breeding programme to introgress the SLICK gene into NZ dairy cattle began in 2014 by crossbreeding dairy cows with Senepol sires. The breeding programme has employed embryo technologies where possible, to accelerate genetic improvement (Breeding Worth, BW) with each generation (Davis et al. 2017).

Cattle carrying the SLICK genotype are characterized by a shorter hair coat (Fig. 1) and demonstrate 0.5-1.0 °C advantage in ruminal and rectal temperature under mild heat load, (THI range 70-76) relative to conventional genetics (Worth et al. 2023). Further, the variation in heat tolerance among individual SLICK cattle is considerably reduced relative to non-SLICK animals (Donkersloot et al. 2021; Worth et al. 2023). An additional benefit of slick cattle in a grazing context, is that slick Holsteins spend more time grazing and directly exposed to solar radiation than their wild-type counterparts (Sánchez-Rodríguez and Domenech-Pérez 2021) when both groups had free-choice to shade.

Coat and skin colour are important to heat tolerance because of the exposure of grazing cattle to the solar heat load and potentially damaging ultraviolet (UV) rays. While slick cattle are inherently heat tolerant it might be expected that additional benefit could accrue in cattle with lighter coats, capable of reflecting a larger proportion of solar load. Black pigmentation is important for the protection of exposed skin against UV rays but a black coat adds to the heat load of grazing animals. Jersey cattle are inherently more heat tolerant than Holsteins (Bryant et al., 2007; Smith et al., 2013). This is likely due, in part, to their lighter coat colour, as well as their smaller size. Causal genetic variation associated with coat colour in cattle has been well-documented (Jivanji 2019) and will ultimately need to become part of any genetic considerations to improve thermotolerance.

Figure 1 The SLICK hair coat in SLICK cattle (A) in summer, relative to that of a non-SLICK contemporary (B)

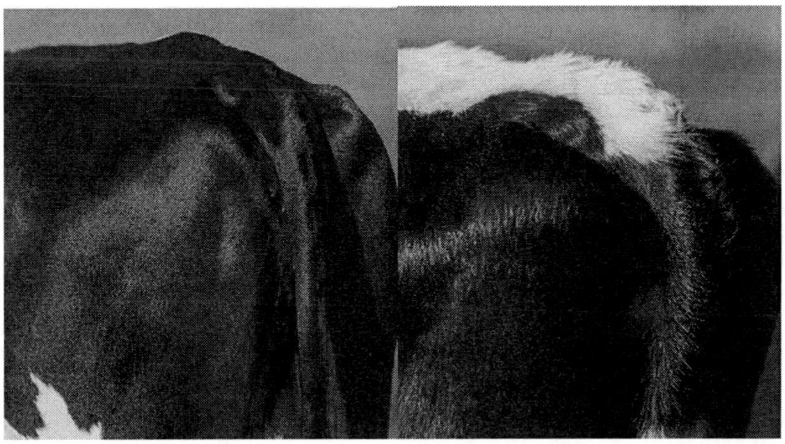

After several generations, sires carrying the SLICK gene have been produced with a BW that is well above the average for the NZ dairy industry and with only 1/32 average content of Senepol genes. BW improvement is ongoing to enhance the likelihood of industry uptake once the genetics are released. Importantly, the SLICK mutation has not been linked to any undesirable dairy traits nor have any undesirable characteristics been associated with the Senepol breed.

Gene editing

Gene editing is now an established technique for the introduction of targeted mutations into the genome and there are number of examples where gene edits for polled, disease resistance or SLICK have been introduced into cattle (Mueller et al. 2019). Approvals for such manipulations have been given by regulatory authorities in USA and Brazil (but not Australia or New Zealand, as of 2023) and are under consideration elsewhere. The advantages of gene editing are several. Firstly, gene editing is much faster than introgression and is potentially much more flexible. Specific mutations like SLICK can be introduced into a high genetic merit background in combination with breed, specific milk protein polymorphisms (A2) and/or traits such as polled. In addition, any other genes that may contribute to variation in hair length/heat tolerance can also be introduced. Achieving such "gene stacking" through introgression is feasible but would be an expensive and very lengthy process, demanding considerable animal resources.

Ultimately a major advantage for gene editing will be to minimise potential inbreeding which is a risk that needs to be managed when attempting to introgress or fix a relatively rare gene in a population (A2 casein variant, Scott et al. 2023).

Conclusion

Cattle carrying the SLICK gene will be part of the global solution for mitigating heat stress in dairying and will have value in New Zealand in providing climate resilience to dairy cattle. Other genetic variants (notably coat colour) and breeding strategies (gene editing) may provide additional advantages. However, such genetic improvements will not obviate the need for management tools such as shade to keep cows cool during heat stress events.

References

Bryant JR López-Villalobos N Pryce JE, Holmes CW, Johnson DL 2007. Quantifying the effect of thermal environment on production traits in three breeds of dairy cattle in New Zealand. New Zealand Journal of Agricultural Research 50: 327-338.

Davis SR, Spelman RJ, Littlejohn MD 2017. Breeding and Genetics Symposium: Breeding heat tolerant dairy cattle: the case for introgression of the "slick" prolactin receptor variant into Bos taurus dairy breeds. Journal of Animal Science 95: 1788-800.

Donkersloot EG, Worth GM, Yeates, AF, Littlejohn, MD, McNaughton LR, Spelman, RJ Davis SR 2021. The benefit of a SLICK hair coat for heat tolerance in New Zealand dairy cattle. Proceedings of the Australasian Association of Animal Breeding and Genetics 24: 94-97.

Garner JB, Douglas M, Williams SR, Wales WJ, Marett LC, DiGiacomo K, Leury BJ, Hayes BJ 2017. Responses of dairy cows to short-term heat stress in controlled-climate chambers. Animal Production Science. 57: 1233-1241.

Harburg S, Amer PR, Duckles J, Jenkins GM, Sise J 2019. Importance of heat stress adaptation for New Zealand dairy cattle. Proceedings of the Australasian Association of Animal Breeding and Genetics 23: 35-38.

Jivanji S. 2022. The genetic characterisation of coat colour and patterning traits in cattle: a thesis presented in partial fulfilment of the requirements for the degree of Doctor of Philosophy in Animal Science at Massey University, AL Rae Centre for Genetics and Animal Breeding, Hamilton, New Zealand. (Doctoral dissertation, Massey University).

Littlejohn MD, Henty KM, Tiplady K, Johnson T, Harland C, Lopdell T, Sherlock RG, Li W, Lukefahr SD, Shanks BC, Garrick DJ, Snell RG, Spelman RJ, Davis SR 2014. Functionally reciprocal mutations of the prolactin signalling pathway define hairy and SLICK cattle. Nature Communications. 5: 5861.

Madalena, FE 1998. Hybrid F1 Bos taurus x Bos indicus dairy cattle production in the State of Minas Gerais, Brazil Proc. 6th World Congress on Genetics Applied to Livestock Production. 27: 199-202.

McDowell RE 1985. Crossbreeding in tropical areas with emphasis on milk, health, and fitness. Journal of Dairy Science. 68: 2418-2435.

Mueller ML, Cole JB, Sonstegard TS, Van Eenennaam AL 2019. Comparison of gene editing versus conventional breeding to introgress the POLLED allele into the US dairy cattle population. Journal of Dairy Science.102: 4215-4226.

Nguyen TT, Bowman PJ, Haile-Mariam M, Pryce JE, Hayes BJ 2016. Genomic selection for tolerance to heat stress in Australian dairy cattle. Journal of Dairy Science. 99: 2849-2462.

Olson TA, Lucena C, Chase Jr CC, Hammond AC 2003. Evidence of a major gene influencing hair length and heat tolerance in Bos taurus cattle. Journal of Animal Science. 81: 80-90.

Pryce JE, Nguyen TT, Cheruiyot EK, Marett L, Garner JB, Haile-Mariam M 2022. Impact of hot weather on animal performance and genetic strategies to minimise the effect. Animal Production Science. 62:726-735.

Porto-Neto LR, Bickhart DM, Landaeta-Hernandez AJ, Utsunomiya YT, Pagan M, Jimenez E, Hansen PJ, Dikmen S, Schroeder SG, Kim ES, Sun J, Crespo E, Amati N, Cole JB, Null DJ, Garcia JF, Reverter A, Barendse W, Sonstegard TS 2018. Convergent evolution of SLICK coat in cattle through truncation mutations in the prolactin receptor. Frontiers in Genetics. 9:57.

Sánchez-Rodríguez HL, Domenech-Pérez K 2021. Light sensors assess solar radiation vs. shade exposure of slick-and wild-type Puerto Rican Holstein cows. J Agric Univ PR. 10539-48.

Scott BA, Haile-Mariam M, MacLeod IM, Xiang R, Pryce JE 2023. Evaluating the potential impact of selection for the A2 milk allele on inbreeding and performance in Australian Holstein cattle. Frontiers in Animal Science 4:1142673.

Smith DL, Smith T, Rude BJ, Ward SH 2013. Comparison of the effects of heat stress on milk and component yields and somatic cell score in Holstein and Jersey cows. Journal of dairy science 96: 3028-33.

Worth GM, Donkersloot, EG, McNaughton, LR, Spelman, RJ, Davis SR 2023. Effects of the SLICK gene on heat tolerance in dairy cattle. New Zealand Journal of Animal Science and Production 83: 5-7

New Zealand Journal of Animal Science and Production 2023 Vol 83.

BRIEF COMMUNICATION: Considerations to cost-effectively automate batch milking in pastoral dairy systems without distributing milkings over 24 h.

JP Edwards[1*], BT Dela Rue[2] & JG Jago[2]

[1]*DairyNZ Ltd, PO Box 85066, Lincoln University, Lincoln 7647,*[2]*DairyNZ Ltd, Private Bag 3221, Hamilton 3240*
*Corresponding author. Email: Paul.Edwards@dairynz.co.nz

Abstract

Current technological limitations mean that to cost-effectively automate batch milking, without distributing milkings over 24 h, a rethink of the farm and milking system is required. A model and a range of non-optimised scenarios were developed for a 300-cow herd to illustrate the trade-offs between the number of automated milking points required (cost) and the duration of 'on-call' time for the person overseeing milking. Compared to the base scenario of 30 milking points (equivalent to conventional milking), by milking the herd in three groups of 100 and adjusting the twice-a-day milking interval from 9-15 h to 6-18 h, the number required could be reduced to 12, with a similar 'on-call' time. Further adjustments could reduce the number of milking points to 7 by milking once-a-day, although five groups of 60 cows would be required to keep cow waiting time below 2 h. These results demonstrate the potential flexibility of current automated milking technology for batch milking.

Keywords: dairy farm systems; automated milking; batch-milking

Introduction

Labour efficiency on New Zealand dairy farms has plateaued at ~150 cows/full time equivalent since 2010/11 (Edwards et al. 2015; DairyNZ 2022). Milking is a time-consuming task, requiring on average 19 h/person/week, excluding herding and cleaning time, at peak lactation when 92% of herds were milked twice-a-day (Edwards et al. 2020). While there are options to adjust the twice-a-day milking interval and/or reduce time spent milking (Edwards et al. 2022), a large proportion of farm labour is occupied for fixed periods of the day, reducing flexibility and restricting the time available for non-milking tasks. Reducing milking frequency is an option but milking would still occupy a large proportion of the workforce for fixed periods in the day. Therefore, automating milking, in particular cluster attachment, could reduce peaks in daily workload and enable an increase in labour efficiency. This could also improve workplace attractiveness by removing a repetitive task, reducing unsociable working hours, and increasing workplace flexibility.

Typically, in automated milking systems, milkings are distributed over a 24 h period and cows voluntarily present for milking (John et al. 2019). This pattern of milking distribution minimises the number of automated milking points required and therefore capital expenditure. This approach has seen limited adoption in pasture-based systems, partly because it introduces complexity in managing voluntary cow traffic and pastures and necessitates someone 'on-call' outside of standard work hours. Milking cows in a group or batch could help alleviate this challenge and may be more easily integrated into existing farm practices.

To automate milking in batches, a robotic arm would be required at each milking stall due to the speed limitations of current automated cluster attachment technology (i.e. costly). Consequently, to cost effectively automate batch-milking a rethink of the farm and milking system is required. To quantify the trade-offs between the number of automated milking points required (cost) and the duration of milking or 'on-call' time (important for workplace attractiveness) a model was developed and the effect of changing aspects of the farm and milking system tested via scenarios.

Materials and methods

A model of milking was developed in Excel (Microsoft, Redmond, WA, USA) and contained 8 input variables that affect the number of milking points or the duration of milking. 1) the duration of the longest milking interval (h), 2) the maximum duration between the start of the first milking and end of the last milking (h), i.e. 'on-call' time, 3) the utilisation of the milking points (% of time, to allow for failed milkings or unoccupied stalls), 4) the size of the milking group(s), 5) the number of milking groups, 6) the time between each milking group (min), 7) the maximum time a group can be at the dairy (h; not including time to/from the paddock) and 8) the maximum group average milk yield in the lactation (kg/cow/day). Note that for a given herd size, the number of groups determines the group size and vice versa. The duration of each milking was proportional to the input milking interval (i.e. milk flow rate was assumed constant). Inputs referred to in the methods hereafter are italicised.

The time that each group was required to be milked within was determined by the minimum of three constraints, i) that the first milking (all groups milked) must be completed before the start time of the second milking in order to maintain the *milking interval* (if twice-a-day milking), ii) the second milking must be finished in time to satisfy the *maximum 'on-call' duration* and maintain the *milking interval* of each group, or iii) the *maximum time a group can be at the dairy* being milked.

The milk yield per cow at each milking was estimated using the *milking interval* and the *maximum group average milk yield*. This was further converted to the time required to milk each cow assuming a maximum milking time was used (Jago et al. 2010). This value was divided by the *utilisation* to determine an average milking point occupancy time and multiplied by the input *group size* to calculate the accumulated milking time. Finally, this was divided by the time the group was required to be milked in (determined above) and rounded up to estimate the number of automated milking points needed.

The model was used to determine the number of automated milking points required under a range of non-optimised scenarios with differing constraints for a 300-cow herd (Table 1), starting from a base scenario that was similar to conventional milking.

Results and discussion

There were an estimated 30 milking points (~NZ$250,000/point) required for the base scenario (A), with one milking group and a maximum time at the dairy of 2 h. Despite the maximum allowable 'on-call' time being set to 12 h, the maximum group milking time constraint of 2 h meant that the actual 'on-call' time was 10.4 h. Increasing the number of groups from one to three (100 cows/group) removed this constraint, allowing the majority of the 12 h period to be utilised and the number of milking points to decrease to 20 (scenario B). Increasing the number of groups to four was not beneficial because each group had to be milked in a shorter duration to maintain the milking interval. To ensure that increasing the number of groups would not create additional workload elsewhere and that people would not be required outside of milking, technology such as virtual herding would need to be investigated. From scenario B, increasing the utilisation of the milking system had a small effect on reducing the number of milking points required to 18 (scenario C). Less 'on-call' time is likely to be more attractive to people on-farm. From scenario B, changing the milking interval to 6-18 h (scenario D) decreased the 'on-call' time to 10.6 h (similar to the base scenario), maximised the duration the milking system was operating and reduced the number of milking points required to 12. The key driver of this was allowing for the second milking to begin immediately after the first milking was finished. Extending scenario D, if the objective was to reduce the 'on-call' time to under 9 h, then a milking interval of 5-19 h and 15 milking points would be required. This illustrates the importance of milking interval to 'on-call' time.

The purpose of scenarios E and F were to minimise the number of milking points further. In both scenarios, the number of milking points required was reduced to 10 (compared to base scenario of 30). The trade-off for this reduction was either an increase in 'on-call' time of 3.3 h or a requirement to reduce peak milk yield by 6 kg/cow/d, for example autumn calving has a flatter lactation curve than spring calving (Jarman et al. 2020). The advantage of the latter approach was a similar 'on-call' time to the base scenario. A variation to scenario F showed that if the reduced peak milk yield was achieved by reducing milking frequency to once-a-day, the number of milking points could be reduced to seven within the same 'on-call' time. However, this would require an increase in the number of groups to five and a reduction in group size to 60 to ensure the maximum group milking time constraint of 2 h was not limiting.

Table 1. Estimated number of milking points and operator day length ('on-call' time) from start of first milking start to end of last milking for a range of scenarios. In all scenarios the maximum duration each group could be at the dairy was 2 h and the time allowed between each group was 10 min.

Scenario name	Base	Increase groups	Increase utilisation	Increase milking interval	Increase day length	Lower peak milk yield
Scenario	A	B	C	D	E	F
Peak milk yield (kg/cow/day)	28	28	28	28	28	22
Milking interval (h)	9-15	9-15	9-15	6-18	8.4-15.6	6-18
Group size (cows)	300	100	100	100	100	100
Number of groups	1	3	3	3	3	3
Milking point utilisation	80%	80%	90%	80%	80%	80%
Maximum 'on-call' time (h)	12	12	12	12	15	12
Required time per group (min)	120	59	59	113	120	113
Milking points required	30	20	18	12	10	10
Actual 'on-call' time (h)	10.4	11.7	11.7	10.6	13.7	10.6
Machine operating time (h)	3.4	6.0	6.0	10.6	11.6	10.6

Overall, these results demonstrate the complex interactions that exist between the model inputs and the tension between the number of milking points required and the length of 'on-call' time. The scenarios are illustrative and are not necessarily optimised. Further work should be conducted to determine farmer preferences for this trade-off, and the impact of stacking complimentary technologies. This analysis demonstrates the flexibility afforded by current automated cluster attachment technology when the farm system design is re-imagined thereby increasing the economic feasibility of batch automatic milking without needing to distribute milkings over 24 h.

Acknowledgements

This study was funded by the dairy farmers of New Zealand through DairyNZ Inc (Hamilton, New Zealand), contract CRS2094 (Frontier Farms).

References

DairyNZ. 2022. DairyNZ Economic Survey 2020-21. https://www.dairynz.co.nz/media/5795678/economic_survey_2020-21_a4_booklet_web.pdf [Accessed 24 May 2023].

Edwards JP, Dela Rue BT, Jago JG 2015. Evaluating rates of technology adoption and milking practices on New Zealand dairy farms. Animal Production Science 55: 702-709. doi 10.1071/AN14065.

Edwards JP, Kuhn-Sherlock B, Dela Rue BT, Eastwood CR 2020. Short communication: Technologies and milking practices that reduce hours of work and increase flexibility through milking efficiency in pasture-based dairy farm systems. Journal of Dairy Science 103: 7172-7179. doi 10.3168/jds.2019-17941.

Edwards JP, Williamson JH, Kuhn-Sherlock B 2022. Improving parlor efficiency in block calving pasture-based dairy systems through the application of a fixed milking time determined by daily milk yield and milking frequency. Journal of Dairy Science 105: 7513-7524. doi 10.3168/jds.2022-21847.

Jago JG, Burke JK, Williamson JH 2010. Effect of automatic cluster remover settings on production, udder health, and milking duration. Journal of Dairy Science 93: 2541-2549. doi 10.3168/jds.2009-2949.

Jarman J, W, M., Kay JK, Neal M, Donaghy DJ, Tozer P 2020. Implications of using an extended lactation to change from a spring-calving to an autumn-calving farm system in South Taranaki. New Zealand Journal of Animal Science and Production 80: 143-149.

John AJ, Freeman MJ, Kerrisk KF, Garcia SC, Clark CEF 2019. Robot utilisation of pasture-based dairy cows with varying levels of milking frequency. Animal 13: 1529-1535. doi 10.1017/S1751731118003117.

BRIEF COMMUNICATION: The Efficacy of Natural Substances to Protect Dietary Glutamine from Ruminal Degradation

LP Carlyon[a], TR Boyce[a], LJ Flanagan[a], BRC Alves[a], K Petrovski[a] & M Caetano[a]*
[a]*The Faculty of Animal and Veterinary Sciences, The University of Adelaide, Roseworthy Campus, Mudla Wirra Road, South Australia, 5371, Australia*
corresponding author. Email: mariana.caetano@adelaide.edu.au

Abstract

In ruminants, the rumen protection of amino acids is required to prevent microbial degradation and increase the portion of undegradable dietary protein; therefore, increasing the total concentration of amino acids absorbed in the small intestine. The conditionally essential amino acid, glutamine, is vital to an array of biological processes throughout the body. The objective of the current study was to validate and optimise the use of two natural substances (NS1 and NS2) to rumen-protect glutamine from microbial degradation. Encapsulated glutamine has been tested in triplicate using an *in vitro* approach. It was concluded that, both natural substances provided protection to glutamine resulting in a 33.83% and 10.66% decrease in ammonia release with NS1 and NS2 respectively.

Keywords: amino acids, in vitro test, protein, rumen microorganisms, ruminants

Introduction

Balancing undegradable-dietary protein (UDP) and rumen-degradable protein (RDP) is crucial to maximise the efficiency of production in ruminants (Haddad et al. 2005; Sahoo & Walli 2008). Therefore, supplementing ruminants with rumen-protected amino acids (AAs) has received attention recently. Rumen protection is required to prevent microbial degradation allowing the supply of specific amino acids and increase the level of UDP. By increasing the UDP portion of the diet, these rumen-protected AAs can bypass the rumen and be absorbed within the small intestine and into circulation (Windmueller & Spaeth 1974; Hanson & Parsons 1997). Even though AAs are heavily involved in many biological functions, and their supplementation to livestock animals demonstrates several benefits, their polarity and hydrophilicity play an important role in defining the success and feasibility to rumen protect a diverse range of AAs.

Glutamine is an uncharged, polar AA because of the R side chain-containing an amine group. It has a pKa (COOH) of 2.17 and a pKa (NH_3^+) of 9.13. Glutamine is a conditionally essential AA and is involved in a multitude of biological processes (Häussinger & Sies 1984). These include immune function, specifically in the pro-inflammatory and some anti-inflammatory response mechanisms. Glutamine is also an important energy source for intestinal cells and aids in the growth and repair of enterocytes. This key AA is a major precursor for glutamate and is vital for foetal growth and survival during pregnancy in sheep (Vaughn et al. 1995). Glutamine has endless applications in optimising production in ruminants and is the chosen AA for this study.

The objective of this study was to determine the efficacy of two cost-effective natural substances (NSs) in protecting glutamine from microbial degradation in the rumen through *in vitro* analysis.

Materials and Methods

The rumen-protected glutamine substances, NS1 and NS2, were prepared under proprietary protocols. The protocols considered labour intensity and preparation time in manufacturing the rumen-protected glutamine to obtain a feasible and cost-effective process in a small-scale setting.

In Vitro Analysis

The rumen fluid was collected from an abattoir (TPL Meat Exports Pty Ltd) and transferred into a pre-warmed thermos. The *in vitro* methodology was adapted from Goering & Van Soest (1970) and Suh et al. (2022) with minor adjustments. The incubated bottles contained a negative control (blank), positive controls (glutamine only, NS1 only, and NS2 only), and the food-grade substance-encapsulated glutamine; all replicated in triplicate. The *in vitro* trial was conducted over a total of 480 minutes and sampling was conducted at 10 time points: 30, 60, 90, 120, 180, 240, 300, 360, 420 and 480 minutes in duplicate.

Ammonium Analysis

Nitrogen levels from the *in vitro* samples were indirectly measured using a colorimetric assay for ammonium determination. For this, 1 mL of each sample was added to microtubes containing 20 μL of 6 N HCl solution and then stored at –20°C prior to ammonium determination. Once defrosted, sample tubes are centrifuged at 1,107 g force or 5 minutes, and an ammonium standard solution (1,000 mg/L ammonium) was used to develop a standard curve. The ammoniacal nitrogen was determined by reacting the sample with sodium-salicylate and sodium nitroprusside. Then samples were exposed to a dichloroisocyanurate solution and the colorimetric detection was measured at 650nm by using the Multiskan plate reader (Bio-Rad Laboratories Inc., xMark™ Microplate Absorbance Spectrophotometer; Hercules, California, USA), 1 hour after all reagents were added (ammonium determination 11732; International Standards ISO 1997).

Statistical Analysis

Descriptive statistics and the area under the curve for each of glutamine-containing combination were calculated in SAS 9.4 using the PROC MEANS function.

Results

Glutamine (non-protected) had a stable release throughout the 480-minute incubation, while NS1 had prolonged efficacy in glutamine protection compared to NS2 (Figure 1). At 360 minutes of incubation, there was no difference between non-protected glutamine and glutamine protected with NS2.

Figure 1. Ammonium concentration (mg/L) of the incubated glutamine (Gln), encapsulated glutamine with natural substance 1 (GlnNS1), and encapsulated glutamine with natural substance 2 (GlnNS2) over 480 minutes *in vitro* incubation (mean ± SEM).

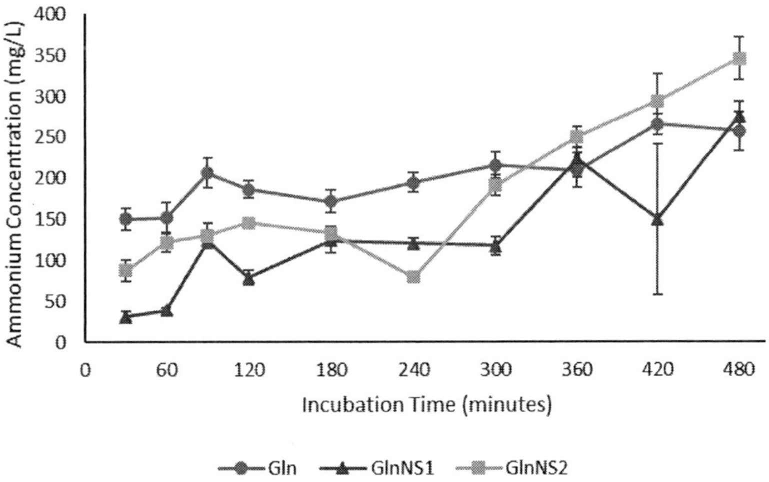

During 480 minutes of fermentation, the glutamine group (non-protected) released 9.24g of ammonia. During incubation, the bottle with glutamine protected with NS1 released 6.12g of ammonia that equates to 33.83% decrease in ammonia release. The bottle with glutamine protected with NS2 released 8.26g of ammonia that equates to 10.66% decrease in ammonia release.

Discussion

The study completed the objective to establish the efficacy of NS1 and NS2 to protect glutamine from microbial degradation in rumen fluid. NS1 resulted in a higher protective efficacy than NS2 as it had a 23.17% less ammonia released during the *in vitro* incubation. Both natural substances were economically feasible to manufacture, requiring minimal preparation and time-effective techniques.

There is dearth of information around the use of rumen-protected glutamine in ruminants and on methodologies and substances used to produce a rumen-protected glutamine supplement. A previous study investigated the use of formaldehyde 2% sprayed on glutamine and dried at room temperature to protect the AA from microbial degradation (Nemati et al. 2018). Depending on the volume of formaldehyde used, methodology is cost-effective being an important factor to consider for future commercial applications. Neither the efficacy nor the bioavailability of this supplement was revealed, therefore conclusions around its protection efficiency cannot be drawn.

Furthermore, current substances used to rumen-protect glutamine usually involve intellectual property and remain undisclosed. A rumen-protected glutamine product made by an Italian company, Ascor Chimici, was used in a study conducted by Caroprese et al. (2012) evaluating glutamine's effect on Friesian cow immune function. The glutamine supplement had a 40% bioavailability. This product is no longer commercially available, bringing into question the rumen protective efficacy and whether the cost of production was feasible. Additionally, rumen-protected glutamine produced by Wansheng Biological Co. Ltd. was used in a study conducted by Wang et al. (2022). This product has a high bioavailability of 65% and on average 25.3% rumen degradability (figures derived from their unpublished research).

The current study measured the efficacy of two natural substances in rumen-protecting glutamine therefore, to determine the bioavailability of the more efficient substance, NS1, an *in vivo* trial would be required. One third of the glutamine intake reaches blood circulation and the residual glutamine is metabolised by intestinal cells (Häussinger & Sies 1984). Therefore, blood glutamine analysis post feeding is a potential measurement for the evaluation of the amount absorbed through the small intestine and available to the body for use. Digesta passage rate and pH would also influence glutamine microbial degradation. Complete degradation of unprotected glutamine in the rumen requires 36 hours (Suh et al. 2022). Different types of feed influence digesta passage rate; high concentrate diets are digested faster than high fibre diets. Differing diets can also influence pH which further impacts rumen microbiome (Vargas et al. 2023). In future studies where the previous diet of the ruminant is unknown, measuring rumen fluid pH would be beneficial.

In conclusion, natural substances 1 and 2 can be used in a cost-effective methodology to prevent partial microbial degradation of glutamine in the rumen. However, further investigation should be performed to evaluate if these natural substances are effective in increasing the levels of glutamine in the small intestine and the blood circulation of ruminants for metabolism.

Acknowledgments

We would like to thank the Davies Livestock Research Centre at the University of Adelaide. Also, thanks to the TSU team and Tong Chen for their help in the ammonium determination analysis.

References

Caropresе M, Albenzio M, Marino R, Santillo A, Sevi A 2012. Immune response and milk production of dairy cows fed graded levels of rumen-protected glutamine. Research in Veterinary Science 93: 202-209.

Goering HK, Van Soest PJ 1970. Forage Fiber Analysis (apparatus, reagents, procedures and some applications), US Agricultural Research Service.

Hanson PJ, Parsons DS 1977. Metabolism and transport of glutamine and glucose in vascularly perfused small intestine rat. Biochemical Journal 166: 509-519.

Häussinger D, Sies H 1984. Glutamine Metabolism in Mammalian Tissues, Berlin, Heidelberg, Springer Berlin / Heidelberg.

International Standard (ISO) 1997. Water quality—determination of ammonium nitrogen by flow analysis (CFA and FIA) and spectrometric detection. Int. Stand. ISO 11732

Nemati MM, Menatian SS, Joz Ghasemi SS, Hooshmandfar RR, Taheri MM, Saifi TT 2018. Effect of protected-glutamine supplementation on performance, milk composition and some blood metabolites in fresh Holstein cows. Iranian journal of Veterinary Research 19: 225-228.

Suh JK, Nejad JG, Lee YS, Kong HS, Lee JS, Lee HG 2022. Effects of L-glutamine supplementation on degradation rate and rumen fermentation characteristics in vitro. Animal Bioscience 35: 422-433.

Vargas JE, López-Ferreras L, Andrés S, Mateos I, Horst EH, López S 2023. Differential Diet and pH Effects on Ruminal Microbiota, Fermentation Pattern and Fatty Acid Hydrogenation in RUSITEC Continuous Cultures. Fermentation (Basel) 9: 320.

Vaughn PR, Lobo C, Battaglia FC, Fennessey PV, Wilkening RB, Meschia G 1995. Glutamine-glutamate exchange between placenta and fetal liver. American Journal of Physiology - Endocrinology and Metabolism 268: 705-711.

Wang S, Wang F, Kong F, Cao Z, Wang W, Yang H, Wang Y, Bi Y, Li S 2022. Effect of Supplementing Different Levels of L-Glutamine on Holstein Calves during Weaning. Antioxidants 11: 542.

Windmueller HG, Spaeth AE 1975. Intestinal metabolism of glutamine and glutamate from the lumen as compared to glutamine from blood. Archives of Biochemistry and Biophysics 171: 662-672.

BRIEF COMMUNICATIONS: Development of a Measurement of Anogenital Distance in Sheep

TR Boyce[a], LP Carlyon[a], KR Petrovski[a], BRC Alves[a], LJ Flanagan[a], & M Caetano[a*]

[a] School of Animal and Veterinary Sciences, The University of Adelaide, Roseworthy Campus, Mudla Wirra Road, South Australia 5371

[*]Corresponding author Email: mariana.caetano@.adelaide.edu.au[*]

Abstract

Fertility is a fundamental phenotypic trait used for achieving optimal production efficiency and, no easily adoptable or precise tools are currently available. In this study, anogenital distances (AGD; anus to clitoris - AC and anus to fourchette - AF) were measured twice from 80 Merino ewes and 80 ewe hoggets to predict fertility. Fertility was measured by evaluating the pregnancy scanning results. Measurements of ADG were performed by using digital callipers and high-resolution images. Overall means of measures were greater for the AF in pregnant ewes compared to dry ewes and no association was found for the AC distance except being shorter for hoggets with callipers. The relationship found between one of the AGD, the AF, and pregnancy status demonstrates the potential to use this trait as a cost-effective method for estimation of fertility that can be easily adopted by producers, particularly in hoggets.

Keywords: breeding, clitoris, imaging, pregnancy status, sheep

Introduction

Anogenital distance (AGD) has been defined as the distance from the centre of the anus to the base of the clitoris and is used as a method of identifying those with potential infertility in humans, pigs, cattle, and rodents (Shourabi et al. 2022, Carrelli et al. 2021, Seyfang et al. 2018, Wainstock et al. 2017, Szenczi et al. 2014). However, to our knowledge, there is no current literature on sheep. The aim of the current study was to estimate the capacity of measurements of anogenital distance (AGD; n=2 types) to predict fertility using pregnancy status of Merino ewes and ewe hoggets.

This study was conducted to evaluate the capacity of the first (AC) and second (AF) anogenital measurements to predict fertility using the pregnancy status of Merino ewes and ewe hoggets. The first objective was to determine whether the AGD measurements varied between pregnancy status. The second objective was to compare the different methods (callipers or image analysis) of measuring the AGD. The third objective was to estimate the repeatability of the same AGD measurement when taken twice per sheep.

Material and methods

Animal handling and measurements

The sheep utilised in this study were from a commercial farm located approximately 15 km from Cummins in rural South Australia (Latitude: -34° 14' 60.00" S Longitude: 135° 42' 59.99" E), under approval by the Animal Ethics Committee (S-2023-001). Eighty ewe hoggets and eighty ewes were randomly selected during pregnancy scanning (65 days after the end of joining) with age, body condition score (BCS) (scale 1 – 5), and pregnancy status recorded (0 or 1) until equal numbers of pregnant and dry per age category were reached. All females were naturally served. All measurements were made during pregnancy scanning. Percentages of pregnant ewes and hoggets were at 150% and 154% respectively.

Experimental design

The AGD measurements included the distance from the anus to the clitoris (AC) and the distance between the centre of the anus to the fourchette (AF). Digital callipers (CS-P8"S; Mitutoyo; Takatsu-ku, Kawasaki, Kanagawa, Japan) with 0.01mm resolution were used to measure each AGD (in mm), twice for each sheep by a single observer. High-resolution images were captured using a prototype device and used to measure AGD with an image analysis software (CMEIAS-IT 1.28; Michigan, America). The calliper and image measurements were mirrored for creating duplicate data.

Statistical analysis

Data analysis was carried out using SAS 9.4 software (Statistical Analysis Software, Cary Inc, USA). The effect of the method used to measure AGD (n = 2; callipers or image analysis) and age of animals (n = 2; ewe hoggets and ewes) on the AGD measurements (AC and AF) including both measurements per sheep was estimated using mixed model in PROC MIXED. BCS had no effect on AGD measurements. The level of significance was set at $P<0.05$. Correlations between measurements of the AGDs between the first and second measures were estimated in PROC CORR.

Results

Means of AF measurements were longer in all pregnant females with callipers ($P = 0.05$) and for ewe hoggets only with image analysis ($P = 0.0001$). However, means of AC measurements were shorter in ewe hoggets with the calliper only ($P = 0.05$; Fig 1) There was a very high correlation between the first and second measurements performed by assessor, being $r=0.98$ ($P<0.0001$) for image or high correlation, being r=0.70 (P<0.0001) for callipers respectively (Table 1).

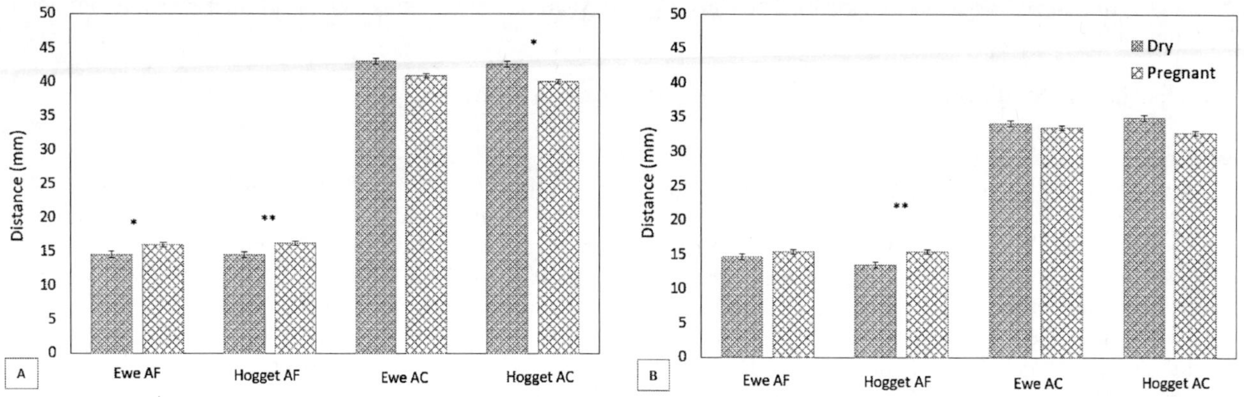

Figure 1: Anogenital measurements (AC = distance from the anus to the clitoris; AF = distance from the anus to the fourchette) between dry and pregnant Merino sheep using callipers (A) and image analysis (B). * P<0.05; ** P<0.0001 between pregnancy groups per age (ewes or hoggets only).

Table 1: Summary table of the AGD (AC = distance from the anus to the clitoris; AF = distance from the anus to the fourchette) values obtained between first and second measurements in mm.

AGD	Measurements	Category	Calliper	Image
			Mean ± SEM	Mean ± SEM
AF	1st	ewe	14.92 ± 1.47	15.21 ± 1.47
AF	1st	hogget	15.36 ± 1.62	14.3 ± 1.61
AC	1st	ewe	37.32 ± 1.45	30.27 ± 1.4
AC	1st	hogget	38.12 ± 1.435	38.12 ± 0.97
AF	2nd	ewe	15.09 ± 1.47	15.27 ± 1.47
AF	2nd	hogget	14.90 ± 1.61	14.34 ± 1.61
AC	2nd	ewe	37.69 ± 1.45	30.48 ± 1.4
AC	2nd	hogget	38.88 ± 1.14	38.71 ± 0.96

Discussion

The comparisons between ewe hoggets and the ewes indicated that in developing females only one of the AGD measurement, the AF, could be used to estimate the fertility (Fig 1), as there was a difference between pregnant and non-pregnant females and methods used. Similar results have been reported for the AGD measurement of the anus to the clitoris in cattle and other species (Carrelli et al. 2021). Thus, the methodology used in this study could be used in a large majority of mammalian species.

An interesting finding from the current study is how the measurements relate to the estimated fertility of the ewe hoggets. The AC distance had minimal difference in pregnant and non-pregnant sheep being shorter only in pregnant hoggets on calliper assessment. In contrast, pregnant hoggets had a longer AF length. This indicated that AGDs used in this study were probably independent. In other studies, the shorter AGD from the anus to the clitoris correlated with better fertility in nulliparous heifers (Rajesh et al. 2022 and Carrelli et al. 2021). Contrary, in Murciano-Granadina goats, aged between one to four years, females with a longer distance between the anus to the clitoris had better reproductive performance (Shourabi et al. 2022).

Based on the results of this study, callipers can be used for AF measurement in all females whilst AC in ewe hoggets only. The image analysis showed an AF measurement being correlated with fertility, but only in ewe hoggets.

The intra-observer correlation between the first and second anogenital measurements of the same AGD was very high (r=0.98) for the image analysis and high (r=0.70) for callipers. This indicated that there is no need for repeated measuring. Notably, calliper to image analysis measurement in ewes had higher variability than in hoggets. Means of calliper measurement were much higher than image analysis in ewes (Table 1). This may have been a result of higher stretch on calliper measurements in ewes and may have skewed the data. As image analysis is based on data capture at one point of time without stretching, seem that is more precise.

Our study had a limited number of observations. Further evaluation of the AGD measurements to estimate fertility with the aim to ensure applicability to external populations is required with a larger sample size in a commercial setting. Furthermore, with the accumulation of data, the use of either a ranking or categorising system can be used to predict fertility. The extra-genetic and genetic factors will need to be evaluated for their effects on fertility and correlations to AGDs. In commercial settings, measuring AGDs to identify fertile ewes is difficult because of changes in anatomical configuration with aging and parturition. Additionally, the fertility of the ewes is already known, and retrospective prediction is fruitless. Therefore, the measurements should be taken earlier in life. Results of this study suggest that this was possible in hoggets after their first joining.

Using the AGDs to estimate the fertility in sheep can allow for the earlier selection of breeding flock. However, there are limitations to using this method in commercial settings. Although minor, additional handling is required that slows down the usual farming practices. Additionally, individual records are required. This may not be feasible for commercial farms. Indeed, it can be feasible for stud farms with smaller flock sizes. Sheep stud farms are more involved with the improvement of their flock and will be more likely to take advantage of using the AGD to estimate the ewe fertility. If they do implement this methodology, it may potentially have a flow-on effect on the commercial farms.

This research showed that the AGD measurement, the AF distance was related to fertility. Based on the higher repeatability of intra-observer measurements, the image analysis method for measuring AF was more accurate and can allow for potential artificial intelligence. Overall, the AGDs measured in this preliminary study had some evidence to be used as predictors for Merino ewe hoggets and, less likely, ewes' fertility.

Acknowledgements

The University of Adelaide, Davies Livestock Research Centre, the Ruminant Nutrition Group, Brad and Kerry Claughton, Cameron Drakow, Jessica Meldrum, Annabelle Hon, Matthew, Rachael, and Acacia Boyce. Without these people the research was not able to happen otherwise and each of them also deserves recognition for the outcome of this study.

References

Carrelli JE, Gobikrushanth M, Corpron M, Rajesh I, Sandberg W, Colazo MG, Ahmadzadeh A, Oba M, Ambrose DJ 2021. Relationship of anogenital distance with fertility in nulliparous holstein heifers. Journal of Dairy Science 104: 8256-8264.

Rajesh I, Gobikrushanth M, Carrelli JE, Oba M, Ambrose DJ 2022. Repeatability of anogenital distance measurements from birth to maturity and at different physiological states in female holstein cattle. Journal of Dairy Science 105: 2699-2707.

Seyfang J, Ralph CR, Hebart ML, Tilbrook AJ, Kirkwood RN 2018. Anogenital distance reflects the sex ratio of a gilt's birth litter and predicts her reproductive success. Journal of Animal Science 96: 3856-3862.

Shourabi E, Hakimi H, Baqeri A, Gharagozlou F, Vojgani M, Foroutannejad M, Baghbanani RH, Mobedi E, Akbarinejad V 2022. Evidence that murciano-granadina does with longer anogenital distance are more fertile and prolific and produce heavier and male-biased litters. Animal Reproduction Science 244: 107047.

Szenczi P, Bánszegi O, Groó Z, Altbäcker V 2014. Correction: Anogenital distance and condition as predictors of litter sex ratio in two mouse species: A study of the house mouse (Mus musculus) and mound-building mouse (Mus spicilegus). PloS One 9.

Wainstock T, Shoham-Vardi I, Sheiner E, Walfisch A 2017. Fertility and anogenital distance in women. Reproductive Toxicology 73: 345-349.

AUTHOR INDEX

New Zealand Society of Animal Production (Inc.) (NZSAP)
PO Box 955
Cambridge, New Zealand
www.nzsap.org.nz

ISBN 978-1-7138-9812-2

9 781713 898122

CURRAN ASSOCIATES INC.
proceedings.com

Seminar des FA Ultraschallprufung 2021

Konventionalle und innovative Anwendungen der Ultraschallprufverfahren

Berichtsband BB 177

Berlin, Germany
10 – 11 November 2021